Abdoul-Azize SAMPEBGO
Joachim Bonkoungou

Assessment of water erosion using the aggregation method

Abdoul-Azize SAMPEBGO
Joachim Bonkoungou

Assessment of water erosion using the aggregation method

Technique for qualitative estimation of water erosion

ScienciaScripts

Imprint
Any brand names and product names mentioned in this book are subject to trademark, brand or patent protection and are trademarks or registered trademarks of their respective holders. The use of brand names, product names, common names, trade names, product descriptions etc. even without a particular marking in this work is in no way to be construed to mean that such names may be regarded as unrestricted in respect of trademark and brand protection legislation and could thus be used by anyone.

Cover image: www.ingimage.com

This book is a translation from the original published under ISBN 978-620-6-72394-3.

Publisher:
Sciencia Scripts
is a trademark of
Dodo Books Indian Ocean Ltd. and OmniScriptum S.R.L publishing group

120 High Road, East Finchley, London, N2 9ED, United Kingdom
Str. Armeneasca 28/1, office 1, Chisinau MD-2012, Republic of Moldova, Europe
Printed at: see last page
ISBN: 978-620-8-15519-3

GUIDE DEVALUATION OF LEVEL OF ERODIBILITY OF THE SOILS BY THERE METHOD AGGREGATION: CASE OF THE IRRIGATION FACILITIES OF BELOW WATERSHED OF NARIARLÉ, BASIN OF NAKANBE AT BURKINA FASO

ABDOUL-AZIZE SAMPEBGO

TABLE OF CONTENTS

DEDICATION

HAS SAMPEBGO IMANDOUDINE,

THANKS

There realization of the here document has summer possible thanks to help And in support of several people and institutions. This is an opportunity for us to express our sincere and deep gratitude to them. First of all, we would like to express our deep gratitude to Éditions Universitaires European (EUE) For THE professionalism and the work remarkable achievement for putting our document online. Thanks to our thesis director, Doctor BONKOUNGOU. THANKS, Doctor For your trust, your availability and your many tips. We also extend our thanks to the teaching staff of the Department of Geography of the NORBERT ZONGO universities of Koudougou and Joseph KI-ZERBO of Ouagadougou, for the training we received. A special mention is made to Professor SOMÉ Yélézouomin Stéphane Corentin for his multifaceted support. Our thanks also go to all the staff of the Environment Institute and of Agricultural Research (INERA), of Desk National Soil Survey (BUNASOLS), to the National Meteorological Agency of Burkina Faso (ANAM BF) for the data that were kindly provided to us. We also express our sincere thanks to all the staff of the Ministry of Agriculture, Animal Resources and Fisheries ; to all the staff of the green economy and climate change MEEA/Burkina Faso; of the National Geology Bureau of Burkina (BUNGEB); to the Regional Directorate of Agriculture, Water Resources, Sanitation and Food Security (DRARH /ASA). Our thanks to all the staff of the General Directorate of Hydraulic Infrastructure (DGIH).Thanks to all the producers of: Wèdbila, Nabazana, Pk25, Monastère. Likewise, we express our gratitude to Mrs. Louré from the communal management of agriculture of Koubri. We born we will know to end without demonstrate our recognitions has the place of all my fellow doctors: OUEDRAOGO Ibrahim, ZOUNDI Mahamadi, ZABRE Gustave, ZAN Amadou, SAVADOGO Boureima, who shared with us the moments of pain and joy of research.

INTRODUCTION

The facilities irrigation systems face climate risks related to to changes climatic. These risks climatic are has the origin of many forms and processes of erosion. According to the United Nations Convention to Combat Desertification (CCD), the degradation of the soils directly touch more of 250 million people and threatens around a billion in more than 100 countries. These people are among the poorest, most marginalized and most vulnerable citizens on the planet. THE political plan (Jarraud, 2005, p. 07). Of data of the UNCCD (2015, p. 05) reveal that erosion is responsible for the loss of about 1/3 of the world's arable land. With climate change, these losses are increasing continuously by more than 10 million hectares per year. Over the last 40 years, 25% of continental land has been experiencing extreme degradation, or is undergoing degradation at an accelerated rate. Regional data from the CORDEX project predict very significant changes in the various climatic parameters. Such that THE precipitation, THE temperatures, THE winds, etc. according to of the studies by A. Azize Sampebgo et al. (2024, p. 01), total annual precipitation will generally experience a decrease in the period 2030-2060 for the scenarios weak (RCP 2.6) And means (RCP 4.5). THE rains daily will be of increasingly intense with averages above 300mm. These future climate data plan of the facilities of more in more eroded with of the consequences very important. The phenomenon of silting up, the disappearance of certain watercourses, the reduction in the retention capacity of hydraulic developments, the considerable drop of there production agricultural, And GOOD others consequences will be observed. It is essential to develop effective methods for assessing the forms and processes of water erosion in light of extreme variations in climatic parameters.

1. DEFINITION OF EROSION

Erosion is the destruction or degradation of soils. Several agents intervene in soil degradation. Of these agents we have man or anthropogenic erosion, winds also called wind erosion and water or hydric erosion. The classification according to the phenomena of the degradation of the soils allows us to distinguish three main levels (Image 1): physical degradation, chemical degradation and biological degradation (Yaméogo, 2021, p. 16). Physical degradation of soil concerns the degradation of the soil structure linked to several phenomena: water erosion, caking, crusting, beating, compaction or increase in relative density, reduction in porosity, soil permeability and wind. Chemical degradation is the degradation linked to the deterioration of the different chemical properties of a soil (acidification).

Picture 1: THE different level of erosion

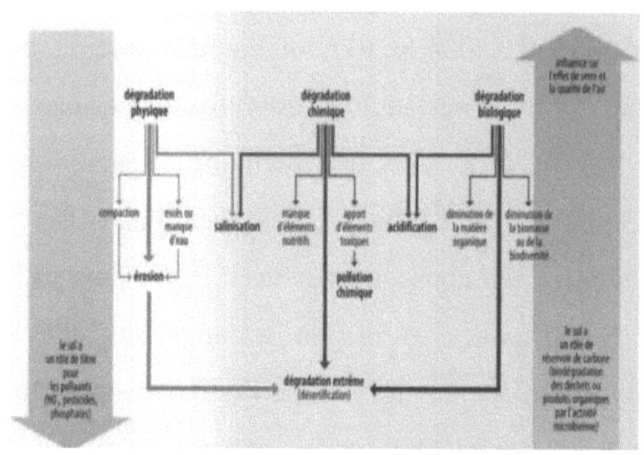

Source : (Back, 2019 quoted by Yaméogo, 2021, p. 16)

Biological degradation is determined by the reduction of the organic matter content of the soil. According to John Agard et al (2014, p. 06), land degradation in arid, semi-arid and dry sub-humid areas refers to the decrease or disappearance of there productivity biological Or economic And East linked of there complexity land. Water erosion processes are nowadays more and more intense with the phenomenon risks climatic (Sampebgo, Ibrahim, And al., 2024, 2024, p. 12).

2. SYNTHESIS OF THE METHODS OF STUDIES OF EROSION WATER

Several methods are used in the study of water erosion. These methods can be grouped into two large groups.

- Qualitative estimation of water erosion and mapping of the terrain according to its vulnerability.
- The quantitative method whose objective is to estimate the quantity of materials evacuated (either according to their points initial, either according to their transportation, either After their deposits).

2.1. THE methods quantitative of erosion water

Three techniques can be used for the estimate qualitative water erosion : THE techniques of measure At level of the slopes, THE techniques of course level measurement water and measurement techniques at the level of lake deposits or reservoirs.

2.1.1. THE techniques of measure At level of the slopes

Slope level measurement techniques. They use four methods.

THE plots experimental , this technical must answer has A certain number of conditions: isolated from neighboring land to avoid interaction, equipped with a basin of reception, THE balance sheet of erosion East established by a analysis of mass volumetric granulometry of sediments (Samake, 2017, p. 50).

Measurements by reference stakes , the progression of erosion processes is measured by stakes, generally graduated iron bars (Feret & Sarrailh, 2005).

There quantification by the radioisotopes ^{137}Cs (cesium 137, 30.2 years of period, depth 25cm has 30cm) And THE ^{210} Lead (22.3 years of period). THE cesium 137 allow to estimate the quantity of soil erosion particles from 1963 (date of maximum 137 $^{Cs\ fallout}$) to the date of analysis of the samples. Gamma spectrometry allows the level of radioactive energy to be measured sampling of profiles and radiation of ^{137}Cs favors the quantification of ^{137}Cs. The operating mode is shown below (Figure 1):

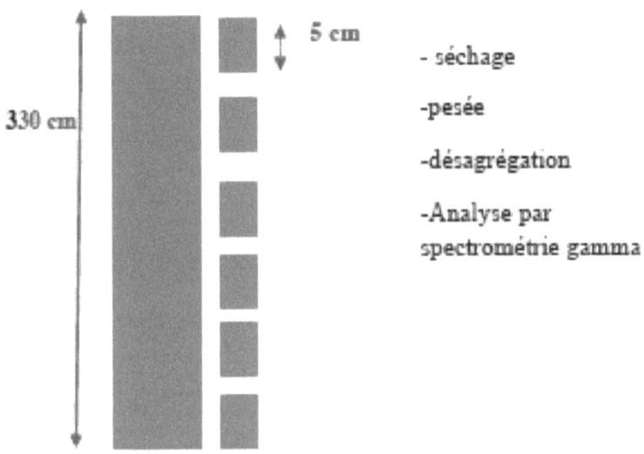

Figure 1: THE fashion operative

THE methods empirical Wishmeier and Smith (1965, 1978) with the USLE formula Or the universal equation of losses in land is present as a formula :

HAS =A. LS. K C. P

HAS : rate annual of losses in earth (ton/hectare/year),

R : climatic aggressiveness or rain erosivity factor (megajoules.mm/hectare. Hour),

LS : topographic factor, he represents the inclination (S in %) and the length of slope (L in m),

K : postman erodibility of the soils (tonne. Hectare. Hour/Megajoules. Hectare. Year),

C : postman of blanket vegetable And of the agricultural practices ,

P : postman of the practices anti-erosive (mulching, cords stony, etc.)

Several authors have considered the empirical estimation of soil erosion at across THE world. There most of these authors to are supported on there formulas Wishmeier

And Smith (1965) In A objective to adapt THE settings of there formula in function study areas. Among these authors we can cite: Knisel (1980) with CREAMS, Foster And Lane (1987) THE WEPP, Morgan (1995) EUROSEM, Fox and Al (1997) with RUSLE, Kinnell in 1998 and Williams and Arnold (2000) with SWAT.

2.1.2. Measuring techniques at the watercourse level and measuring techniques at the level of lake deposits or reservoirs

Measurement techniques at the watercourse level . Transport in solution or suspension and transport by the bottom are the techniques applicable for the qualitative estimation of water erosion at the watercourse level (Abir, 2013; Descroix et al., 1997, p. 05; Mazour & Roose, 1996, p. 05).

Measurement techniques at the level of lake or reservoir deposits . The bathymetry method can be applied to quantify lake or reservoir deposits (Moukhchane, 2002; Rampon, 1990; Sabir, 1986a, p. 150).

2.2. THE methods qualitative of erosion water

Several qualitative methods are distinguished for the description of erosion forms. water. Among these methods we can cite: there description of the forms of erosion, erosion potential, there simulation of rain, THE photographs aerial and remote sensing, magnetic susceptibility of soil particles, PAP/CAR guidelines and aggregation method.

2.2.1. There description of the forms of erosion

There description of the forms of erosion is made by the description of there geomorphological map. The geomorphological map allows to distinguish the different geomorphological of the area of study : THE structures lithological, according to THE types of rocks allow to determine the types and the level of alteration of types of rocks represented. Tectonic, morphological data and symbols determining THE different types of erosion (Map 1). The study diachronic allow of follow the evolution of structures in time and space.

2.2.2. Erosion potential

Erosion potential is based on the overlay of the cartographies themes of factors of erosion and in adding the codes. As examples of factors from erosion we can cite lithology, slope, land use, aggressiveness of precipitation, etc. THE level of erodibility of the materials of alteration East class of 1 to 5 depending on the types of rocks and their durability. The slopes are classified from 1 to 5, 1 for THE slopes null Or very weak And 5 For THE slopes very high (superiors) has 35%). This method presents of the boundaries related At coding of the classifications Who to limit has an arithmetic sequence and the interdependence of factors (Adamou & Maîga, 2011).

2.2.3. There simulation of rain And THE photographs aerial And there remote sensing

There simulation of rain . She analysis THE behavior of ground according to of the simulation methods. This technique is more reliable for the comparative study of reactions of different kind of soils. She laid of issue of extrapolation on of big surfaces (BRGM & Bufalo, 1990; Stéphanie, 1994).

Aerial photography and remote sensing . It allows the diachronic analysis of the evolution of erosion forms, the movement of sediments, cartographic of the factors of erosion (Dubucq, 1986; Laganier, 1994, p. 18; Paul-Hus, 2011; Puech et al., 2006; Raclot et al., 2005).

2.2.4. There susceptibility magnetic of the particles of ground

This technique is applicable in areas with a Mediterranean climate or contrasting seasons using the fersiallitization method. It is based on the magnetic properties of iron to assess the level of soil erosion. The samples are subjected has there susceptibility MS2B, At field magnetic **H** (H= 0.1mT ; millitesla) And has the needle magnetized **Mr.** There curve of susceptibility magnetic

$$(x = \frac{M}{H}, x = 10^{-8} m^3/kg)$$

10^{-8} m^3/kg) depending on the depth (cm) allows to measure the state of erosion between several areas (Figure 2).

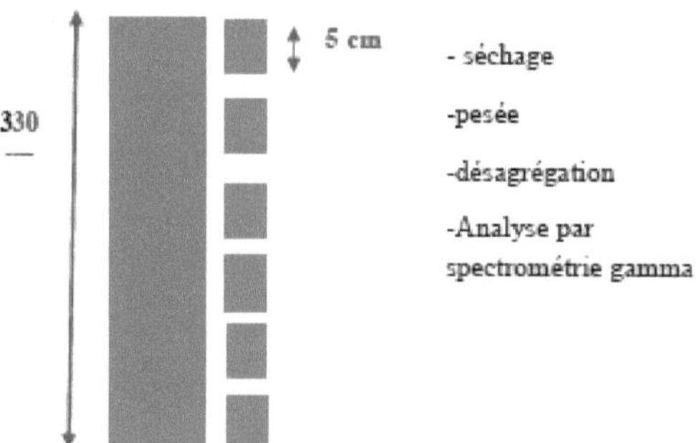

Figure 2: Fashion operative

2.2.5. The PAP/CAR (Priority Action Programme of the Regional Activity Centre for the Mediterranean region) guidelines

It consists of three phases which are: a predictive phase which allows to complete the map of erosion states, a descriptive phase which is a mapping of the forms of real erosions existing and an integration phase which allows to develop the consolidated erosion map according to PAP/CAR. The map of erosion states results from the superposition of there map of erodibility and there protection of ground by there vegetation according to well-defined codes (Table 1, 2).

Painting 1: Class of slopes

Code	Class of the altitudes :749m- 125m
1	Very weak Or null 100- 200
2	Weak 200- 300
3	Average 300- 400
4	Pupil 400- 700
5	Extreme 700- 800

SAMPEBGO HAS. HAS, 2024

1: very low or no water flow, 5: extreme flow accelerated by the slope which is a major factor.

Painting 2: Class litho- facies

Litho-facies class (Code-litho)	Kind of material
1	Rocks compact, No altered, strongly cemented, outcrop of sandstone, limestone
2	Rocks Or soils cohesive, fractured, moderately altered
3	Rocks sedimentary, clayey, THE soils are strongly Or moderately compact
4	Rocks little resistant, strongly altered
5	Sediments Or rocks furniture, No cohesive, materials detrital

SAMPEBGO HAS. HAS, 2024

The soil erodibility matrix (PAP/CAR) (table) allows the layout of the erodibility map into five classes (Table 3).

Painting 3: Matrix of erodibility of soils (PAP/CAR)

Slope class (GID-CODE)	Litho-facies (Code- litho)				
	1	2	3	4	5
1	1	1	1	1	2
2	1	1	2	3	3
3	2	2	3	4	4
4	3	3	4	5	5
5	4	4	5	5	5

SAMPEBGO HAS. HAS, 2024

For a low slope of code 1 the implementation movement of particles only occur on very fine and non-cohesive type 5 materials, whereas for a steep slope the erosion affect by the same intensity the litho-facies of code 4 and 5 (Table 4).

Painting 4: Degrees of erodibility of soils (PAP/CAR)

Classes : (Code Erodibility)	Degrees of erodibility
1	Weak
2	Moderate
3	Average
4	Forte
5	Extreme

SAMPEBGO HAS. HAS, 2024

The protection of the ground by vegetation depends on the nature of land use and of there density of recovery. There class occupation of ground East organized in six (06) classes as noted In THE painting 5.

Painting 5: Class code occupation of the soils

Class : (Occupation code)	Occupation of the soils
1	Bare land
2	Culture
3	Housing
4	Tree-lined or shrubby
5	Gallery forest
6	Forest clear

SAMPEBGO HAS. HAS, 2024

The map of density of vegetation cover of the watershed is composed of four (04) classes (Table 6).

Painting 6: Code covered vegetable

Classes : (Code-cover)	Degree of covered vegetable :
1	Lower has 25%
2	25% - 50%
3	50%- 75%
4	Superior has 75%

SAMPEBGO HAS. HAS, 2024

The application of the soil protection matrix allows the creation of the soil protection map in five classes from very low to very high (Table 7, 8).

Painting 7: Matrix of protection soils

Occupation of ground : (Degree-occupation)	Degree of plant cover : (Code-(to cover)			
	1	2	3	4
1	5	5	4	4
2	5	5	4	3
3	3	2	1	1
4	4	3	2	1
5	5	4	3	2
6	5	4	3	2

SAMPEBGO HAS. HAS, 2024

Painting 8: Classes of degree of protection of the soils

Classes : PAP/CAR	Degree of protection of the soils
1	Very pupil
2	Pupil
3	AVERAGE
4	Weak
5	Very weak

SAMPEBGO HAS. HAS, 2024

The application of the matrix erosive states of soils allows to establish the state map erosive in five class of erosion very weak has erosion very pupil (Painting 9, 10).

Painting 9: Matrix of the state erosive of the soils

Matrix of the state erosive of the soils					
Degree of protection of the soils :	Degree of erodibility :				
	1	2	3	4	5
1	1	1	1	2	2
2	1	1	2	3	4
3	1	2	3	4	4
4	2	3	3	5	5
5	2	3	4	5	5

SAMPEBGO HAS. HAS, 2024

Painting 10: Codification of the states erosive

Classes : PAP/CAR	Degree of the states erosive
1	Erosion very weak
2	Erosion weak
3	Erosion notable
4	Erosion pupil
5	Erosion very pupil

SAMPEBGO HAS. HAS, 2024

The consolidated erosion map according to PAP/CAR is based on the mapping of the erosion state map and the existing real erosion map. Sheet erosion, superficial gullies (rills, ravines), deep gullies (ravines), and solifluction movements.

3. ASSESSMENT OF WATER EROSION BY THE AGGREGATION METHOD: CASE OF BELOW BASIN SLOPE OF NARIARLE, BASIN OF NAKANBE AT BURKINA FASO

3.1. Presentation of there area of study

The Nariarlé watershed is located at latitude 12°12'54.03" north and longitude 1°19'46.57" west of Burkina Faso (Map 2). It is defined as a geographical entity global And coherent For a management of there resource in water (Boismartel, 2012, p. 06). The Nariarlé watershed is a sub-basin of the Nakanbé (one of the national basins of Burkina Faso). It is limited to the East by the Nakanbé, to the West by the Nazinon, to the North by the Massili and to the South by the Pendjarie. The Nariarlé watershed covers seven (07) communes: Four from the central region (Koubri, Saaba, Komsilga and Ouagadougou) and three from the central-south region, from the province of Bazèga (Saponi, Kombissiri and Boulgou).

Map 1 : Situation of there area of study

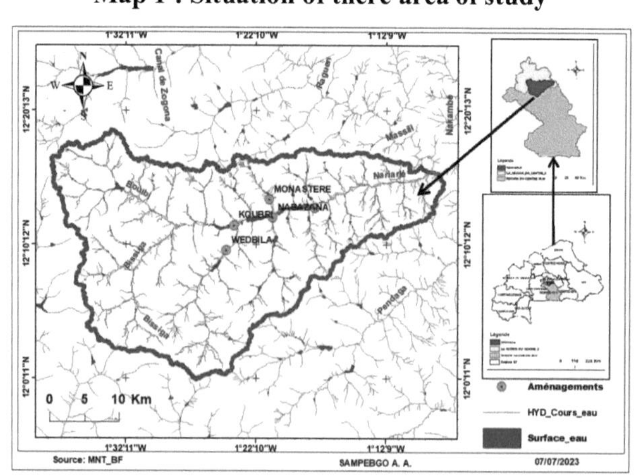

3.2. Methodology

The assessment of the shapes and of the process of erosion by the aggregation method is based on the fifth IPCC report. The technique requires four methods: The structure of an impact chain, therefore the objective is to **determine and prioritize the erodibility factors** (Figure 3). It can be organized around the following factors according to the objectives of the study:

- Postman of protection of the soils
- Factors topographical (the inclination Or there length of there slope).
- THE factors climatic (THE hazards climate Or THE settings extreme climate).
- THE factors anthropogenic such that THE practices anti- erosive.
- THE factors pedological.

There method of standardization. THE indicators are standardized in using there

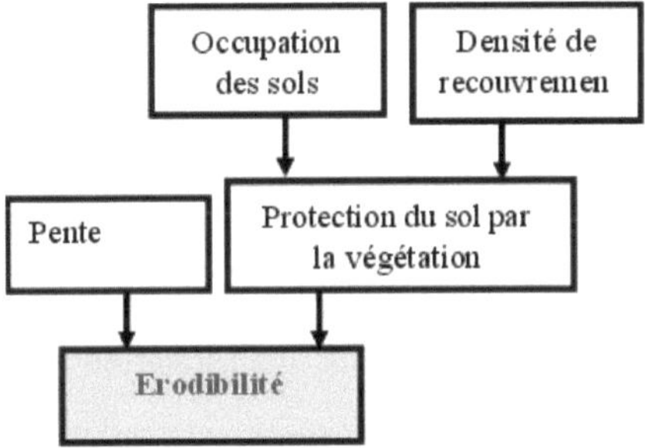

Figure 3: Structure of the arithmetic aggregation chain of erosion in the Nariarlé sub-watershed

$$Tn_{i,0/1} = \frac{X_i - X_{min}}{X_{max} - X_{min}} \quad \textbf{Eq. 1}$$

The aggregation method. It allows a comparative study of the level of erodibility. Aggregation is defined as a combination of information from different indicators in A indicator composite (IC) below there shape of a component unique (GIZ, 2021, p. 122).

$$IC = \frac{(I_1 \times W_1 + I_2 \times W_2 + I_3 \times W_3)}{\sum_1^n W} \quad \textbf{Eq. 2}$$

W East THE coefficient assigned has the indicator. Mapping of real erosion forms and processes by GIS (Google Earth)

THE network hydrographic East the whole of the course of water natural Or artificial, permanent or temporary which drain the waters of a watershed towards the outlet (Yaméogo, 2021, p. 86). Water East mon of the main agents climatic of erosion irrigation developments in the Nariarlé watershed. The characterization of the hydrographic network of the Nariarlé watershed is generated by the Digital Terrain Model (DTM). The parameters used are: the area (S), the perimeter (P), the hierarchy, the index of Gravelius also called index of shape (Kg), there density of drainage (Dd).These parameters are determined automatically using QGIS software. The characteristics physical characteristics of the Nariarlé sub-watershed are obtained by remote sensing. It is a scientific discipline that has made it possible to bring together all the knowledge and techniques for the observation, analysis, interpretation and management of the environment of the sub-basin, from measurements and images (DEM, sentinel 2) obtained using the Copernicus/open hub download platform. Remote sensing allows the acquisition of information remotely, without direct contact with the object studied (Bendjoudi & Hubert, 2002, p. 14; Chamaillé, 2008, p. 10; Forsing et al., 2008, p. 03; Ozer & Karimoune, 2009, p. 06). The most discriminating spectral bands are selected to facilitate and optimize the classification. SPOT satellites can detect objects of about ten meters in size on each of the 60-kilometer-wide images. Images acquired by SPOT 5 (launched THE May 4th 2002 by the rocket Ariane 4) are used For the development of the detailed land use and vegetation cover map. The National Topographic Database (BNDT) made it possible to identify localities, administrative boundaries, road networks, water surfaces and gullies of irrigation developments (IGB, 2015). The types of irrigation development correspond to the Nariarlé sub-watershed, the outlet points of which were generated from the database of the regional director of agriculture and hydraulic developments and technical data from the ONBAH design office (DRAHDI/MAAH, 2020).

3.3. THE fashion of determination And of calculation hydrological

• There area And THE perimeter of watershed

The delimitation and calculation of the area(s) of the watershed were done using ARCGIS software from the DTM. The perimeter of the watershed (P) represents there length of there line of sharing of the waters delimiting THE basin (Baguemzanre, 2007, p. 28). He has summer determined has leave of there delimitation on ARCGIS software from Burkina SRMT MNT.

• The index of Gravelius Or hint of shape kg

It is a clue Who allow of define there shape of the watershed. Indeed, if Kg is equal has 1, we have to do at A watershed circular, otherwise Kg > 1 and all the higher as the watershed is elongated (Eq. 1).

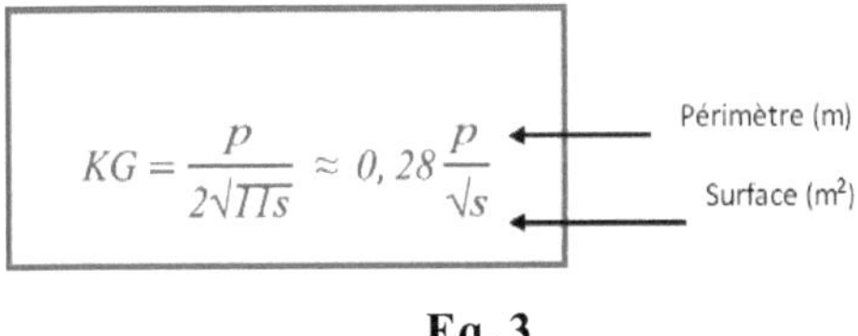

Eq. 3

- **There length (L) And there width (L) of watershed**

$$L = \frac{p + \sqrt{p^2 - 16s}}{4}$$

Eq. 4

- P: perimeter (m)
- S : surface (m^2)

$$l = KG\frac{\sqrt{S}}{1,128}(1 - \sqrt{1 - (\frac{1.128}{KG}})$$

Eq. 5

- **There density of drainage (Dd)**

It is defined as the ratio between the total length of the hydrographic network and the surface area of the watershed. It gives an indication of the importance runoff across the watershed. The total length of streams draining the watershed was measured using ARCGIS software.

- **There slope average (I)**

The slope average East data by there formula next :

$$I = \frac{\Delta H_{max}}{L_{cours\ d'eau}}$$

Eq. 6

P average : slope average of course of water [m/km] ;

Δ H max : maximum river elevation [m] (difference in altitude between the furthest point and the outlet);

L : length of course of water main [km].

3.4. THE methods delimitation of the facilities irrigation

The sub-basins of Nariarlé (Boussouma, PK25, Wédbila, Monastère, Nabazana) correspond to the types of irrigation facilities. The sub-basins and their hydrographic networks are created automatically by the QGIS software. The watercourses are classified autonomously according to the Strahler method:

- All course of water not having not tributary East said of order 1

- To confluences of two course of waters of even order " **n** " , THE course water result is of order " n + 1 "

- A stream receiving a lower order tributary retains its order THE points outlets of the below basins correspond to contact details geographical of the facilities hydraulics of the perimeters irrigated (Painting 12).

Painting 11: contact details of the outlets according to THE sites

Points outlet	Latitude	Longitude
KOUBRI PK 25	12.19	- 1.39
WEDBILA	12.16	- 1.40
NABAZANA	12.20	- 1.34
MONASTERY	12.22	- 1.35
BOUSSOUMA	12.21	- 1.29

Source : DRAH/HAAH SAMPEBGO A. A., 2023

There Water control is one of the determining elements for securing agricultural production. This control requires a good knowledge of hydrological regimes

And more particularly THE features of the floods exceptional And low water levels. The exceptional contributions of runoff water reorganize the hydrographic network

3.5. There hierarchy of network hydrographic of basin slope of Nariarle

The organization of the hydrographic network of the Nariarlé watershed is automatically generated by the software ARCGIS using the Strahler method (1952). According to the types of organization of river channels of Derruau (1974) the basin slope East of type channels anastomosed. THE network hydrographic a A dendritic aspect, It is a set of course of water branched, the more frequent in a normal uniform erosion environment, "the dendritic type corresponds either to uniformly resistant sediments, horizontal or beveled by a horizontal surface, or to crystalline rocks whose slope is generally slight". The hierarchy of the hydrographic network of the Nariarlé watershed (map 3).

Map 2: the organization of network hydrographic of basin slope of Nariarle

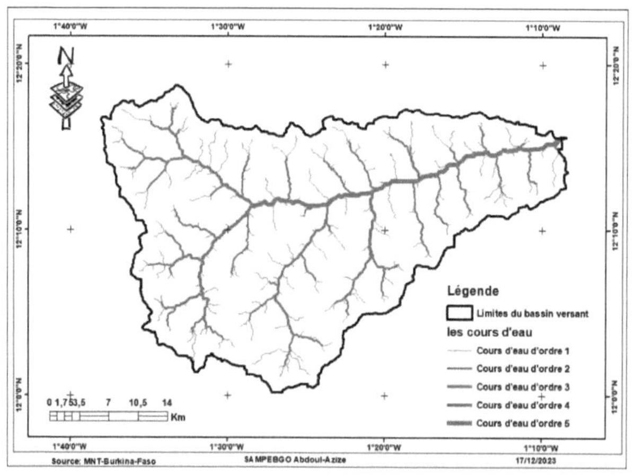

There shape of network hydrographic can be give by of the empirical formulas of report of confluence $RC = \frac{N_n}{N_{n+1}}$ Eq. 2) And THE report of RL length $RL = \frac{N_n}{N_{n-1}}$ (Eq. 3) named THE laws of Horton. THE results of these two clues to present

In THE painting below (Painting 13).

Painting 12: THE results of the reports of confluence And of length of watershed

Order of the course of water	Number of course of water (L n)	THE report of confluence (RC)	Length (Li ; Km)	THE report of length (R li)
1	357	1.63	404.15	2.6
2	219	3.65	154.86	2.77
3	60	1.13	55.85	1.79
4	53		31.18	0.74
5	109	1.81	42	
THE	798	2.5 < King < 5	688.04	

SAMPEBGO TO. TO., 2023

THE basin East of kind circular, A network dendritic kind oak with of the well-developed branching and regular spacing of confluences.

Painting 13: features hydrological of basin slope

Features		Values	Units
Geographic coordinates	X	681,733	Km
	Y	1350.88	Km
Area	HAS	1060.71	Km 2
Perimeter	P	209	Km
Length of the course of water main	L	404.15	Km
Altitudes	Maximum : Z max	356	m
	Minimal : Z min	257	m
	Elevation gain : Δz	99	m
Slope : ΔZ L	S	0.24	
Density of drainage	∑ n L Li Dd = i=HAS	0.64	Km /km 2
Hint of compactness	I understand = 0.282. P. S - 1/2	1,809	
Rectangle equivalent	L = S 1/2 . (I comp / 1,128). [1 + (1 - (1,128/ I comp) 2) 1/2]	93.04	Km
Hint overall of slope	H 5 % – H 95 % Ig =The eq	0.72	m/km
Elevation gain specific	D S=Ig . √ A	24.10	m

SAMPEBGO HAS. A. 2023

THE results observed are : low dynamism and runoff of the basin which is explained by the geology, topographical characteristics, climatological and geomorphological conditions of the basin, The drainage density of the Nariarlé watershed is 0.64. It expresses the ratio of the total length of permanent and temporary watercourses (Li=688 km) to the surface area of the watershed (A=1060.7139 km ^{2}), THE number total of the course of water of basin slope of Nariarle (L n) East of 798 on an area of 1060.7 km2 · The hydrographic density of the basin corresponds to 0.75 ^{km2} . These results testify that THE basin slope of Nariarle rest on a very permeable parent rock, little vegetation cover and little relief (Table 14).

3.6. Erodibility of below watershed

THE vulnerabilities has erosion of below watershed East deduced by there rate of alteration or friability of the basin and protection of the soil by vegetation.

3.6.1. There slope

The higher the slope of the basin, the more it creates unfavorable conditions for irrigation and vice versa. (Table 15).

Painting 14: : Standardization of the altitudes

THE irrigation facilities	Altitude	Average altitude	Normalized value	Vulnerability level
Boussouma 1	267-356	311.5	0.67	Weak
Monastery	277-323	300	0	Very weak
Nabazana	270-356	313	0.76	Average
Koubri	273-356	314.5	0.85	Pupil
Wèdbila	280-354	317	1	Very pupil

SAMPEBGO HAS. HAS., 2023

The results of the study reveal that the highest altitudes are the most developed of Wèdbila, PK25. Nabazana records a altitude average, the one Boussouma and the Monastery represent the weakest.

3.6.2. There protection from the ground by there vegetation

Soil protection by vegetation depends on the nature of land use and the coverage density. The soil protection map is produced by superimposing the land use map and the coverage density map.

- **Map occupation of ground**

Land use map reveals strong demographic pressure on resources natural features of irrigation schemes (Map 4). The comparative analysis of the land use map from 2017-2022 shows an extension of built-up areas 40.14%, an increase in the rate of soil degradation of 37.35%, an increase in the irrigated agricultural area of 45.51%. In eight (08) years, irrigation developments have experienced deforestation of 108 ha (08.21%) and a reduction in water reservoirs of 484.8 ha (23.07%), reflecting the disappearance of certain watercourses in the developments. The expansion of the rural population, the increase in the needs of families, impoverishment, low acquisition of efficient irrigation tools and techniques have an impact on the increase in climatic risks.

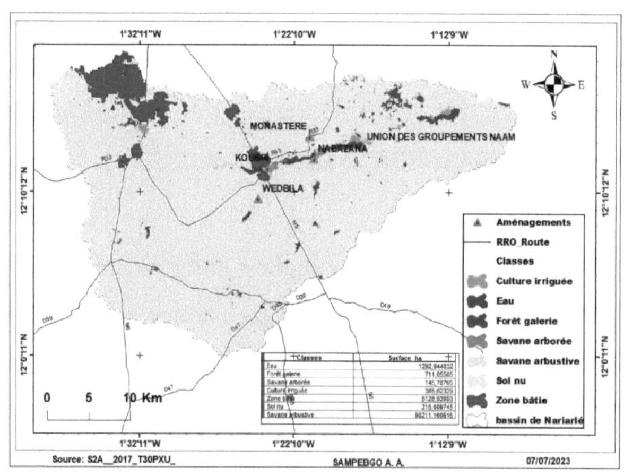

Classes	Surface_ha
Eau	1292,944032
Forêt galerie	711,05565
Savane arborée	145,78765
Culture irriguée	385,62329
Zone bâtie	5128,93093
Sol nu	215,009745
Savane arbustive	90211,160616

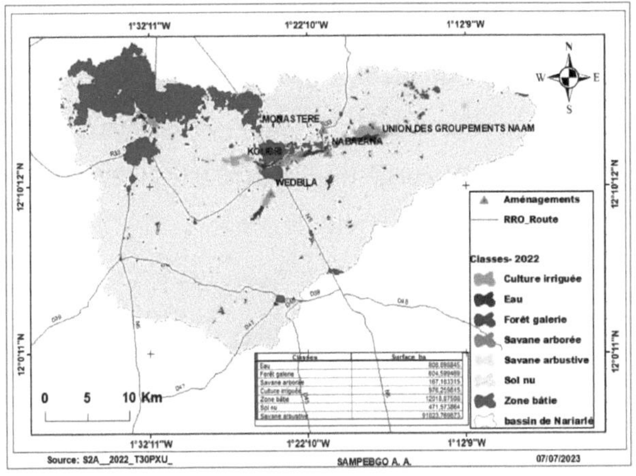

Map 3: Occupation of the soils

The level of land occupation of the different irrigation schemes is determined according to the area occupied by the classes or occupation indicators. value standardized of the facilities East obtained in calculating there average of the min-max of indicators (Table 16).

Painting 15 : THE level occupation of ground

THE irrigation facilities	Normalized value	Occupancy level
Boussouma 1	0.169	Average
Monastery	0.148	Weak
Nabazana	0.176	Pupil
Koubri	0.204	Very pupil
Wedbila	0.145	Very weak

SAMPEBGO AA, 2023

- **There density of recovery Or rate of plant cover**

THE below watershed of Nariarlé is a area of savannah by Excellency (Map 5). The types of vegetation encountered: the gallery forest which occupies 16.6% of the basin (16.6 %), there savannah tree-lined (32.08 %), there savannah shrubby (28.3 %), there savannah grassy (23%).

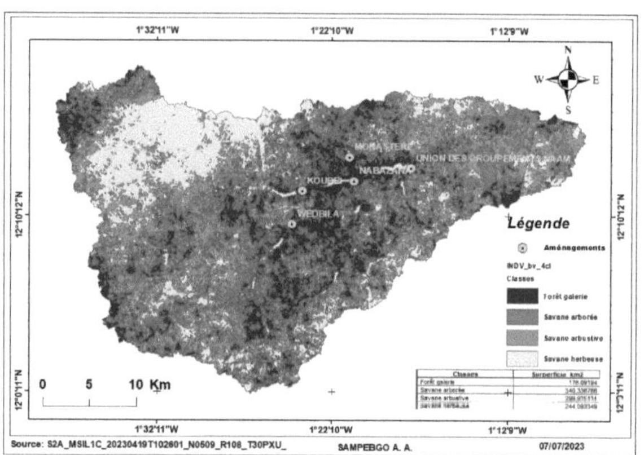

Map 4: Vegetation cover of the Nariarlé sub-watershed

Table 16: Aggregation of the densities recovery by coefficient

THE sites	S	Density of recovery				Agreed value	Land use (Normalized value)	Value of vulnerability (inverse normalized)	NV
		Forest gallery and	Sava not tree-lined	Savanna and shrubive	Savanna and grass use				
		4	3	2	1				
Boussoumy 1	854	0.69	0.9	0.26	0.18	0.19	0.16	0.84	Pupil
Monasterye	21.67	0.8	1.41	0.4	0.05	0.24	1	0	Very weak
Nabazana	679	0.60	0.84	0.52	0.28	0.2	0.33	0.67	Averageenne
Koubri	450,1	0.39	0.72	0.54	0.246	0.18	0	1	Very pupil
Wedbila	148,6	0.76	1.11	0.56	0.14	0.23	0.83	0.17	Weake

SAMPEBGO AA, 2023

S : Area (in km ^{2})

NV : Level of vulnerability

The results show that the Monastery and Wèdbila developments have a high occupation efficiency with a high coverage density. They are less susceptible to climate hazards (Table 17). The soil protection map is obtained by superimposing the occupation map and the vegetation cover map using the aggregation method (Table 18).

Painting 17: Aggregation of there map of protection of ground

Irrigation facilities	Land use	Plant cover	Aggregate value	Vulnerability level of floor protection
Boussouma 1	0.16	0.84	0.5	Pupil
Monastery	0.14	0	0.07	Very weak
Nabazana	0.17	0.67	0.42	Average
Koubri	0.20	1	0.6	Very pupil
Wedbila	0.14	0.17	0.15	Weak

SAMPEBGO AA, 2023

3.6.3. There map of erodibility of the facilities irrigation

The level of erodibility of irrigation schemes (Map 6) is determined statistically by aggregation of the values standardized of there slope And of protection of soil (Table 19).

Painting 18 : THE level of erodibility of the irrigation facilities

Irrigation facilities	Normalized value and of the slope	Normalized value e floor protection	Value of erodibility	Vulnerability level of erodibility
Boussouma 1	0.67	0.5	0.58	Average
Monastery	0	0.07	0.03	Very weak
Nabazana	0.76	0.42	0.59	Pupil
Koubri	0.85	0.6	0.72	Very pupil
Wedbila	1	0.15	0.57	Weak

SAMPEBGO AA, 2023

The level of erodibility remains high for the Nabazana and Koubri developments. It is average for the Boussouma developments and respectively low and very low for the Wédbila and Monastère developments.

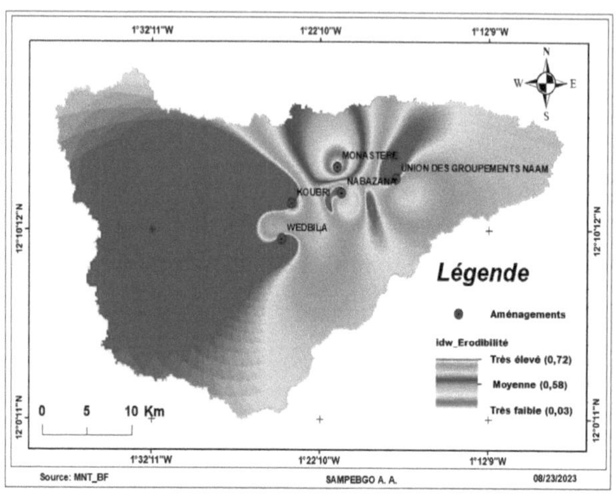

Map 5 : Erodibility of the facilities irrigation of below watershed of Nariarle

3.7. Forms And process of erosion real

THE facilities irrigation of basin slope of Nariarle are confronted to the different process of erosion related to hazards climatic. These process are has the origin of the actual forms of erosion observed in the basin (Map 7): sheet, rill, ravines and **erosion by watercourses** . Water is one of the main climatic agents of erosion of the irrigation facilities of the Nariarlé watershed. Erosion water of the facilities begin by the impact of the drops of rain on the ground, these are the first movements of soil particles. Raindrops crash into the ground, digging a crater and injecting soil constituents several centimeters all around: this is the splash effect. The distance traveled depends on there size of there droplet, of the angle of incidence Thus that of the inclination of there slope of there surface of ground. Energy of the drops of the rains do explode THE aggregates And show them structures of ground. She scattered THE particles And there turbulence provokes by the impact of the hammering provokes a destruction, a reorganization And a redistribution particles fines between the particles coarse. THE particles fines are trained in the porosities and deposited in the micro depressions. This process of soil aggressiveness causes there bet in place of a crust clayey called crust of beating between the plots of arrangements. There crust of beating clayey trains THE soil runoff which will begin to erode and transport particles.

Map 6: erosion real of watershed of Nariarle

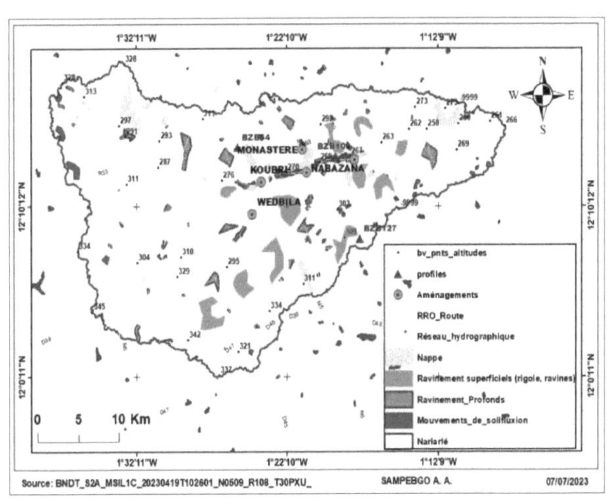

Runoff carries more soil particles on a clayey crust than on a surface with intact aggregates. Runoff gradually strips the soil surface: **this is sheet erosion** . It depends on the slope, the intensity of the duration of rainfall, the structural stability of the soil and the resistance of the particles to erosion. Erosion in tablecloth also called erosion pellicular is an incised erosion and marks an intense erosion of the watershed. It impoverishes the soils in fine and nutritious elements for irrigated crops that are struggling to grow. Stripping the soil also denies tree roots, which can be uprooted under the action of wind (Picture 2). She East located at the level of the plains At ground hydromorphic and pseudogley, At round of the fields of but to altitudes 680327.57 E And 1350700.45 N (BZB 111) ; of the glaze At ground ferruginous tropical laundry indurated deep below wooded and clear savannah, located at 679796.83 E and 1337316.28 N (BZB 170).

Picture 2: uprooting of the trees

Sheet erosion is also observable on the lower slope of eutrophic brown soil of the peanut fields (map 8) located at 695499.5 E and 1355149.22 (BZB 85) And on THE glaze superior of 659572.19 E And 1349715.94 (BZB 043) characterized by a shallow, hardened, leached tropical ferruginous soil of millet fields under shrubby savannah with rocky outcrops (cuirass).

Map 7: location of the profiles

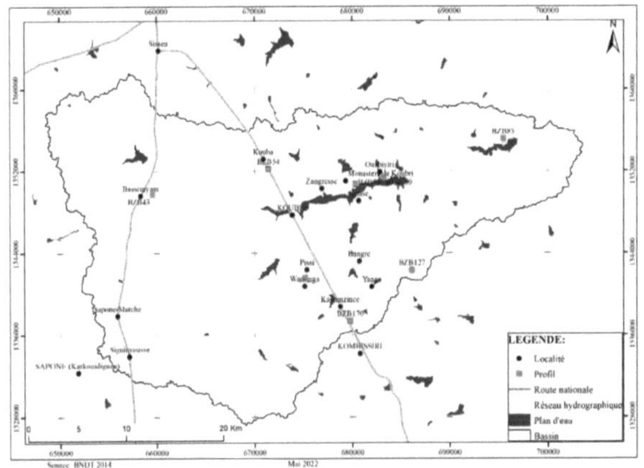

SAMPEBGO AA 02/06/2022

Rill erosion is the first form of incision and linear digging of soils (Image 3). It occurs when there are irregularities on the soil surface, often due to cultivation methods (Sabir, 1986b, p. 10). It is located at the level of the middle and lower slopes of tropical ferruginous soils leached with stain and concretion, almost flat with moderate drainage (BZB 127, 109,). Several types of rills are observed depending on the form of incision and the types of soils: Single-horse channels : they have isolated incisions and are mainly linked to the topography. A scion to creates following a depression lying down Or on of the cultivated land under ploughing at contour lines. Parallel channels : they have parallel incisions, they are observable on of the lands plowed following there slope. THE scions left by THE The ploughing will serve as a drainage channel for the water which will continue to dig them. Branched or hairy ditches : they are observed on naturally cultivated land. Branched or hairy ditches are generally erased after plowing. They play a very important role in the soil loss and transport of nutrient particles from irrigated soil.

Picture 3 : erosion in a gutter

SAMPEBGO AA 02/06/2022

When the runoff in the rills intensifies, they become a flow and acquire of the energy that him allow of dig more deeply notably on the floors weakened And worked by THE labor. The evolution of the gutters do to be born of the ravines. These ravines are also observed following a deep dissection of the soil creating a sudden break in the slope. The channels evolve into ravines depending on the slope, under the effect of the intensity of successive rains (Image 4).

Picture 4 : ravines

SAMPEBGO AA 02/06/2022

The study of the watershed shows the presence of ravines. These ravines are due to the evolution of channels and gullies. The flow of water makes them increase in depth and width.

Picture 5: erosion by the course of waters

SAMPEBGO AA 02/06/2022

Erosion by watercourses. It manifests itself by a brazenness of the bank of the shore concave And A sliding of there bank of there convex bank. Erosion by the most important watercourses irrigation developments are observed upstream And in downstream of the dams. That testifies of the shortcomings of the studies of construction of these water reservoirs. Erosion is also done by ablation of the funds according to the capacity of the course of water has create Or has carry its charge of sediments. She depends of there speed of the current and its energy (Image 5). The study of the real erodibilities of the Nariarlé watershed in Koubri shows that the irrigation schemes in Burkina Faso are subject to water erosion. Erosion water has summer identified as mon of the elements majors of there degradation of soils cultivable of the basin slopes (Avakoudjo And al., 2015, p. 02; Dugue, 2007, p. 37; YAAGOUB et al., 2016, p. 43). According to (Duchemin et al., 2004, p. 31) the rate of soil erosion will change in response to climate changes. It negatively affects THE attributes functional pores of transmission And of conservation water And constitute an obstacle major For the achievement of self-sufficiency eating of the growing world population (Pimentel, 2006). The erosivity of rainfall in the Rif Western knows a evolution (CHOUCHRI, 2016, p. 12), according to the climate projections produced by four Global Climate Models of two scenarios. According to (Douaoui & AL, 2004, p. 8) the massive supply of non-saline water during the first rain contributes to leaching the soluble salts and thus to very quickly reducing the electrical conductivity of the soil solution. Two mechanisms intervene in the process of erodibility of the soils of the irrigation facilities in Koubri: the phenomenon of detachment of the elements of land by the impact drops of rain on the ground and the phenomenon of detachment. These results are in accordance with the work of Dumas J. (1965, p. 03). The integration of geographic information systems (SIG) has the analysis of the factors involved of erosion presents many advantages, Above all those related At big number of results relative (Tribak And al., 2012, p. 06). There method , despite his boundaries, brought a help important to decision makers to simulate of the scenarios devolution And plan the interventions of struggle against erosion (Yjjou et al., 2014, p. 08).

CONCLUSION

This study was carried out as part of a thesis document entitled " risks linked to climate change in irrigation developments in the Nariarlé sub-watershed in Koubri, Nakanbé basin, Burkina Faso " which aims to devalue THE risks climatic of the facilities irrigation of below Nariarlé watershed in Koubri in Burkina Faso according to IPCC report five. Irrigation developments At Burkina Faso are confronted with risks of erosion related to climate change. The extreme variation in climatic parameters intensifies the exposure of irrigation facilities to erodibility risks. The study made it possible to know the forms and real erosion processes of the facilities in Koubri. The implementation of effective practices and techniques for adaptation to climatic risks are the best means of combating water erosion of irrigation facilities both in Burkina Faso and in Africa.

REFERENCES

Abir, BS (2013). Role of ravine erosion in the siltation of hill reservoirs in the Tunisian Dorsal and Cap Bon [Continental Waters and Society]. National Agronomic Institute of Tunisia (INAT).

Adamou, MM, & Maîga, O. (2011). Measuring potential runoff and erosion in the Boubon watershed (middle course of the Niger River) . 04.

J. , & Schipper, ELF (2014). Annex II, IPCC Glossary (p. 201). http://www.ipcc.ch/site/assets/uploads/2018/02/AR5_WGII_glossary_EN.pdf

Avakoudjo, J., Kouelo, HAS. F., Kindomihou, V., Ambouta, K., & Sinsin, B. (2015). Effect of erosion water on THE features physicochemical of ground of the erosion zones (dongas) in the Commune of Karimama in Benin. African Agronomy ,27 (2), 127 - 143.

Baguemzanré, KSD (2007). Development studies of the Tô lowland in the province of Sissili (p. 94) [Memory for obtaining of [diploma in rural engineering]. 2iE.

Bendjoudi, H., & Hubert, P. (2002). THE coefficient of compactness of Gravelius: Critical analysis of a watershed shape index. Hydrological Sciences Journal ,47 (6), 921 - 930. https://doi.org/10.1080/02626660209493000

Woodmartel, Mr. (2012). Situation of there area of study .31.

BRGM, & Bufalo, M. (1990). Rain simulation application to the study of infiltration and erosion processes of agricultural land in Lauragais (Hôte Garonne) .44.

Chamaillé, L. (2008). Contribution of remote sensing for water detection and characterization of water properties (p. 78) [Master 2 professional geomatics. « Georeferenced Information Science for Environmental Control and Regional

Planning » (SIGMA)]. Toulouse Academy.

UNCCD. (2015). Climate change and land degradation : Connecting knowledge to issues . 19.

Descroix, L., Guedez, P.-Y., & Poulenard, J. (1997). Methods of measure of current erosion : Applications in the southern Pre-Alps (France), and the Sierra Madre Occidental (Mexico) .

Dubucq, Mr. (1986). Remote sensing spatial And erosion of the soils : Study bibliographical .12.
Duchemin, Mr., Rousseau, HAS. N., Majdoub, R., & Quilbe, A. (2004). Impacts potential effects of climate change on soil water erosion . 37 (04), 32.

Dugue, P. (2007). Erosion And his mechanisms. Burkina Faso , 37.
Dumas, J. (1965). Relationship between soil erodibility and their analytical characteristics. Orstom Notebooks , 307 - 333.

Feret, J.-B., & Sarrailh, J.-M. (2005). Use of a simple, and accurate, measuring device for the study of erosion in Mayotte .
https://agritrop.cirad.fr/528996/1/document_528996.pdf

Forsing, J.-M., Roux, E., & Ose, K. (2008). fllVilTllllenlelltt geographi[lue : Contributions from remote sensing, mapping and GIS .07.

GIZ. (2021). Guide complementary on there vulnerability : The concept of risk (p. 68). Jarraud, M. (2005). Climate and land degradation . WMO-No. 989 , 34.
Laganier, R. (1994). Contribution to the study of erosion processes and natural risks in the islands of the southwest Pacific (New Caledonia and Solomon Islands) [Geography].

Mazour, M., & Roose, E. (1996). Influence of vegetation cover on runoff and soil erosion on erosion plots in watersheds of northwestern Algeria . 17.

Moukhchane, Mr. (2002). Different methods of estimation of erosion In THE watershed of Nakhla (Rif Western, Morocco). Bulletin of NETWORK EROSION , 21 , 266.

Ozer, T. HAS. HAS., & Karimoune, S. (2009). Contributions from remote sensing in the study of the environmental dynamics of the Tchago region (north-west of Gouré, Niger) Contribution of Remote Sensing in the study of the environment dynamic in the region of Tchago (north-west of Gouré, Niger) (p. 12). https://geoecotrop.be/uploads/publications/pub_331_06.pdf

Paul-Hus, C. (2011). Methods for studying erosion and management of degraded sites in New Caledonia .

Puech, C., Scrapper, D., & Jacome, HAS. (2006). Of the images of remote sensing For erosion.04.

Raclot, D., Puech, C., Mathys, N., Roux, B., Jacome, A., Asseline, J., & Bailly, J.-S. (2005). Aerial photographs taken by drone and Digital Terrain Model : Contributions to the Draix erosion observatory. Geomorphology : relief, processes, environment , 11 (1), 7 - 20. https://doi.org/10.4000/geomorphologie.209

Crawl, HAS. (1990). Water erosion and sedimentation In THE dams. CEMAGREF ,78 , 08.

Sabir, M. (1986a). Water erosion and its quantification .PARIS XI Department of HYDROLOGY and ISOTOPIC GEOCHEMISTRY.

Sabir, Mr. (1986b). Erosion water And its quantification [Memory of DEA].University of Paris xi department of hydrology And of geochemistry isotopic. Samake, O. (2017). Estimation of soil erosion below cultivation in zone Sudanese from Mali : Case of the village of Kani (Koutiala circle). [For obtaining the Master's degree in Agronomic Sciences from the IPR/IFRA of Katibougou. Specialty : Integrated Soil Fertility Management (ISFM).].

https://cgspace.cgiar.org/server/api/core/bitstreams/593a13dd-1186-4c73-9710-1f7ad47349de/content

Sampebgo, A.-A., Ibrahim, O., & Joachim, B. (2024). Climate Risks of Irrigation Developments in the Nariarle Sub- watershed in Koubri, Nakanbé Basin, Burkina Faso. Asian Journal of Advances in Agricultural Research, 24(5), 50-64. https://doi.org/10.9734/ajaar/2024/v24i5506

Sampebgo, A.-A., Zan, A., & Bonkoungou, J. (2024). Forms and processes of real erosions of the facilities irrigation of below watershed of Nariarle, basin of the Nakanbe At Burkina Faso. Review International of there Research Scientist and of Innovation(Review-IRSI) ,2 (2),Article 2.https://doi.org/10.5281/zenodo.11243668

Stéphanie, B. (1994). Study of irrigation by rain simulation of three Mediterranean vineyard soils . 149.

Tribak, HAS., The Garouani, HAS., & Abahrour, Mr. (2012). Erosion water In THE Tertiary marl series of the eastern peninsula : Agents, processes and quantitative evaluation. Review Moroccan of the Science Agronomic And Veterinarians ,1 (1), 47 - 52.

YAAGOUB, D., SOUILAH, K., JAAFARI, Y., Raouf, P. J., Abdel-Ali, P. C., &Mohammed, P. B. (2016). Memory of END of studies .43.

Yaméogo, A. (2021). Characterization of erosive dynamics in the upper watershed of there Sissili (Burkina Faso) [Thesis of [PhD]. University Joseph KI-ZERBO.

Yjjou, M., Bouabid, R., El Hmaidi, A., Essahlaoui, A., & El Abassi, M. (2014). Modeling of water erosion via GIS and the universal equation of soil losses in the Oum Er-Rbia watershed. The International Journal Of Engineering And Science (IJES) ,3 (8), 83 - 91.

CLAUSE OF DISCLAIMER (ARTIFICIAL INTELLIGENCE)

The author declared by there presents that none technology of AI generative such that major language models (ChatGPT, COPILOT, etc.) and text-to-image generators were not used when writing or editing manuscripts.

INTERESTS COMPETITORS

The authors have declared that he does not exist not of competing interests .

AUTHOR (S)

Sampebgo Abdoul-Azize

Laboratory of Research in Science Humans, Department of Geography, Norbert Zongo University, UFR/SH, BP: 376, Koudougou, Burkina Faso.

Research And experience academic :

Environment And development of territory. Dynamics-Space and Society.

Specialization : *GIS, Cartography, Geomatics, climate risk assessment in many sectors such as agriculture, environment, water, sanitation, soil health, etc.*

***Bonkoungou Joachim** (Researcher, Researcher senior)*

INERA/CNRST, National Research Center scientist And technological/Institute of Environment and Agricultural Research, Burkina Faso.

***Research and academic experience** : He is a senior scientist in geography, climate And environment. It has supervised of the research practices of students PhD and Master II researchers.*

***Specialization in research** : Her domain of specialization understand mainly the evaluation of the risks climatic In of many areas. Of the sectors such as agriculture, environment, water and sanitation, soil health, etc.*

Others) points) notable(s) : He East member from Africa Climate Leadership.

Printed by Books on Demand GmbH, Norderstedt / Germany